S0-ESD-819

MILESTONES

James A. Warner and Margaret J. White

MILESTONES

A MIDDLE ATLANTIC PRESS BOOK

Copyright © 1989 by James A. Warner and Margaret J. White

All rights reserved. No part of this publication may be reproduced or transmitted in any form or by any means, electronic or mechanical, including photocopy, recording, or any information storage and retrieval system, without permission in writing from the publisher.

First Middle Atlantic Press printing, November 1989

The Middle Atlantic Press, Inc.
848 Church Street
Wilmington, Delaware 19899

Library of Congress Cataloging-in-Publication Data

Warner, James A.
  Milestones.

  1. Photography—Portraits.   I. White, Margaret J.
II. Title.
TR680.W37   1988   779′.2′0922   88-13620
ISBN 0-912608-62-5

Distributed by:
National Book Network, Inc.
4720 A Boston Way
Lanham, Maryland 20706

Printed in Hong Kong by Everbest Printing Company through
Four Colour Imports, Limited, Louisville, KY.

JAMES A. WARNER

The Gentle People, A Portrait of the Amish
The Quiet Land
Songs That Made America
The Darker Brother
The Mormon Way
Best Friends
The Journey
Chesapeake, A Portrait of the Bay Country
The Decoy as Art, Waterfowl in a Wooden Soul
In the Footsteps of the Artist,
Thoreau and the World of Andrew Wyeth
Shakespeare's Flowers
Rites of Passage

MARGARET J. WHITE

Best Friends
The Journey
Chesapeake, A Portrait of the Bay Country
The Decoy as Art, Waterfowl in a Wooden Soul
In the Footsteps of the Artist,
Thoreau and the World of Andrew Wyeth
Shakespeare's Flowers
Rites of Passage

Text references in *Milestones* are made to the following books:

The Gentle People, A Portrait of the Amish, 1985, The Middle Atlantic Press, Wilmington, Delaware (6th printing); James A. Warner and Donald M. Denlinger.

The Quiet Land, 1988, The Middle Atlantic Press, Wilmington, Delaware (2nd printing); James A. Warner.

The Darker Brother, 1974, A Dutton Visual Book, E. P. Dutton & Co., New York; James A. Warner and Styne M. Slade.

The Mormon Way, 1976, Prentice-Hall Inc., Englewood Cliffs, New Jersey; James A. Warner and Styne M. Slade.

Best Friends, 1980, A & W Publishers, New York; James A. Warner and Margaret J. White.

The Journey, 1980, Osmond Publishers, Salt Lake City, Utah; James A. Warner and Margaret J. White.

Chesapeake, A Portrait of the Bay Country, 1982, Creative Resources, Winterville, North Carolina; James A. Warner and Margaret J. White.

Milestones was begun in the Fall of 1980 as a study in the portraiture of civic, cultural, personal, religious and social rites that are integrated into the rhythm of our lives.

Almost nine months into the project, the photographers put the work aside to publish portrait essays on four subjects: the Chesapeake Bay; decoys as works of art; the countryside of Chadds Ford, Pennsylvania, home of American artist Andrew Wyeth; and the flowers and flowering herbs mentioned in the plays, poems, and sonnets of William Shakespeare.

In 1987, work on *Milestones* was resumed. The finished product presents some of the most thought-provoking, powerful images to come from the photographic studio and gallery of renowned photographer James A. Warner and his distinguished associate of ten years, Margaret J. White.

Milestones

This is a penetrating, engaging view of life's big and small moments as seen through the eyes of the photographers. Here is humor and tenderness and captivating serenity held in images you will not soon forget.

Milestones is a journey into moments with—and without—proper names through the fine art of portraiture.

Photographers' Note

References in the text to portraits done in the studio and to studio and gallery collections of prints refer to the Warner Studio and Gallery in Forest Hill, Maryland. The studio is the center for the exhibition and sale of portraits and books by James A. Warner and Margaret J. White.

Milestones

A FIRST PUPPY

**When you give a child a dog, you don't just give him
four little feet, a cold, wet nose with a tiny, waggly tail
—you give him a whole new life.**

The sentiment and the portrait are from the photographers' 1980 portrait-essay about the relationship between people and their dogs titled, *Best Friends.* The mother of the little girl in the picture raised Sheltie dogs and had brought three pups to the studio for a portrait of them before they were sold. This moment between the girl and one of the pups was too precious to forget.

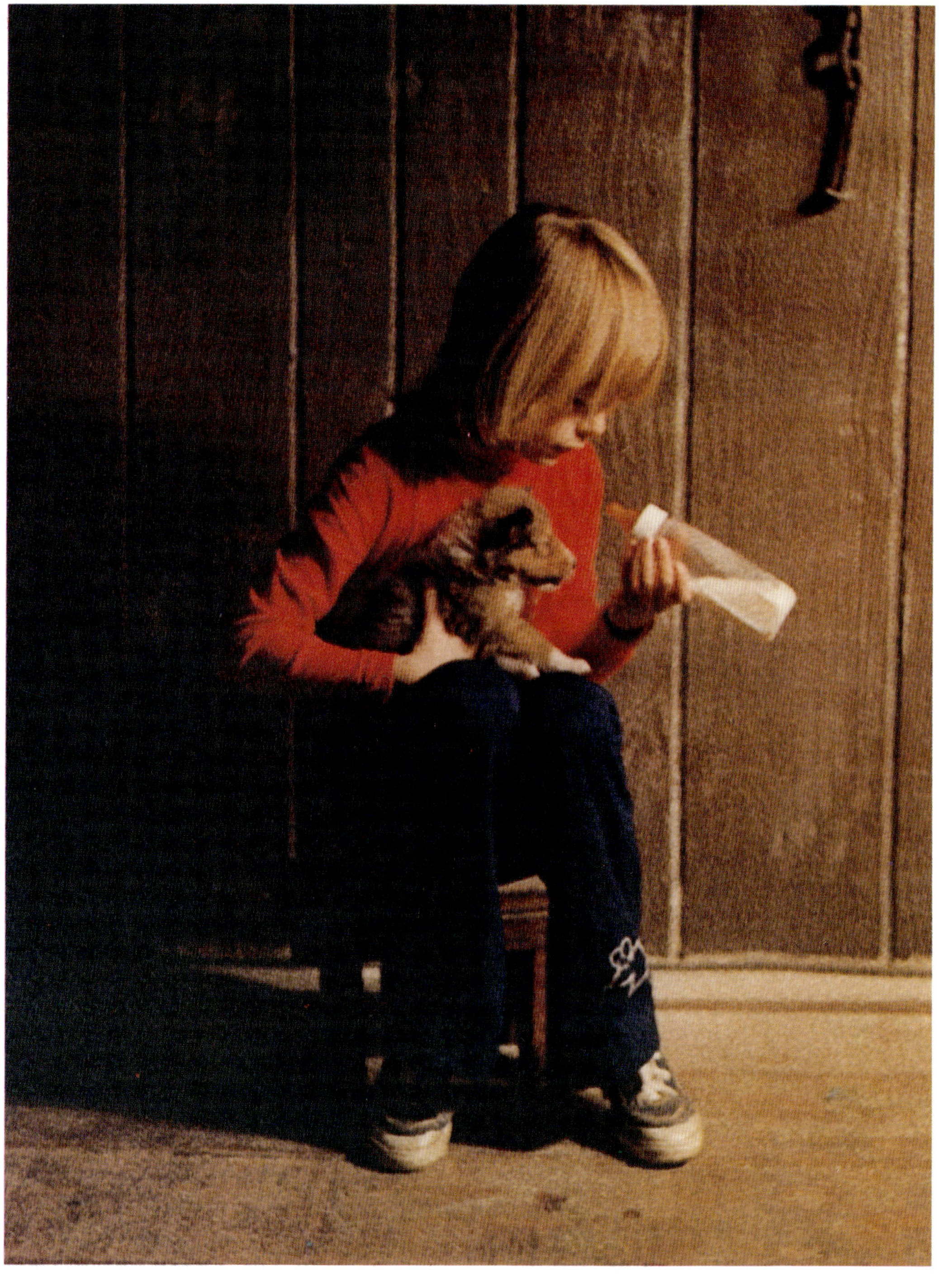

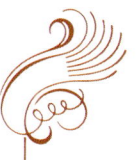

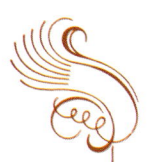

A FIRST VALENTINE

The studio was used for the setting of this elementary school scene. The exchange of valentines during a class party was the inspiration for the portrait. The young boy playing hard-to-get is Margaret White's son; the little girl caught in the motion of a shy, backward glance is his Sunday School friend.

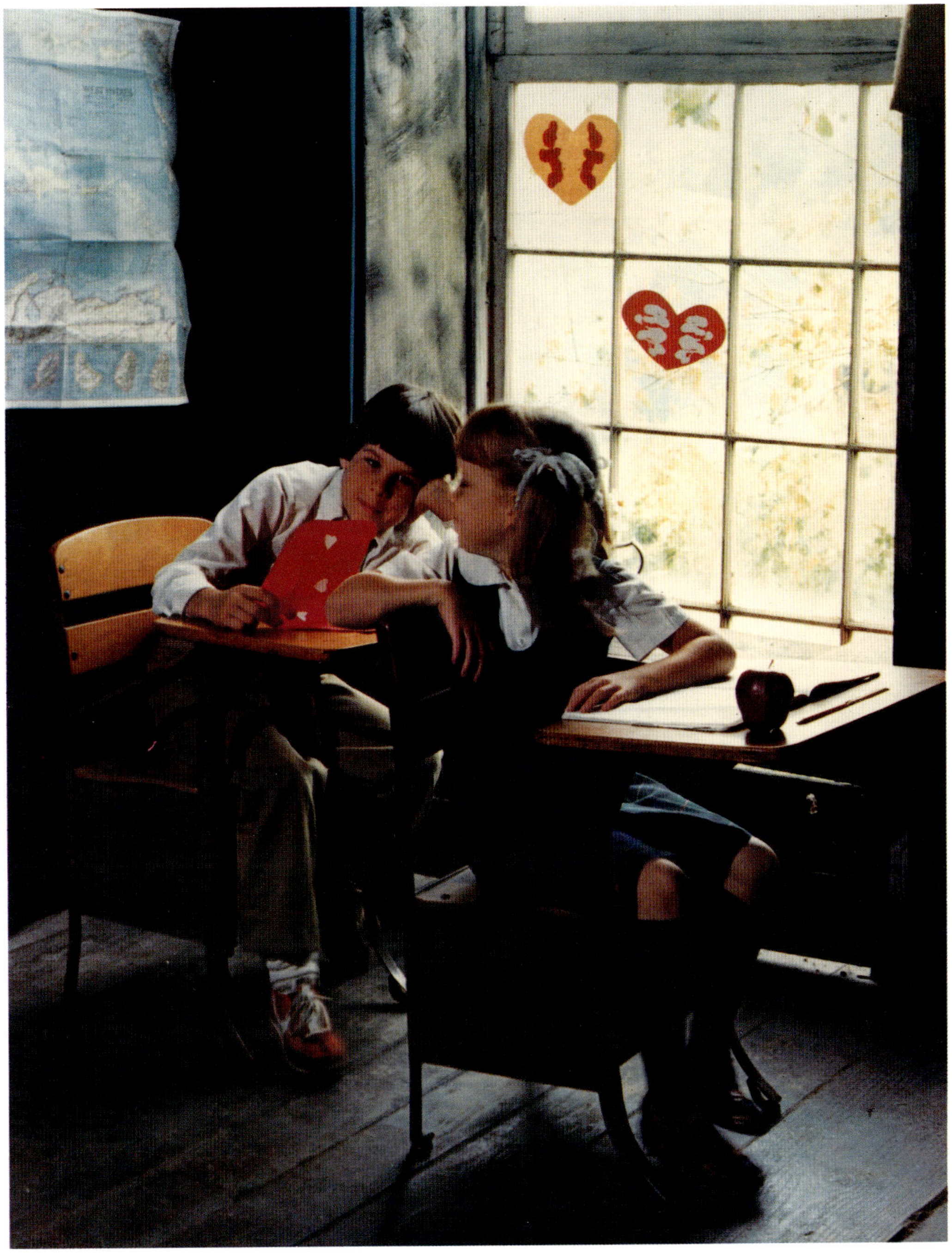

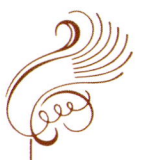

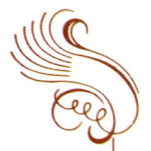

Grandfather and Grandson

While setting up for this scene at the studio, Margaret White was recalling a moment that had passed between her father and his first grandchild.

"I'm your grandfather. Oh yes I am!" he'd say each time he took his infant grandson into his arms. One day she asked him why he felt the need to introduce himself each time he talked to the baby. "So that he'll always remember who I am," he replied.

The same conversation seems to be taking place between the grandfather and grandson in this studio portrait.

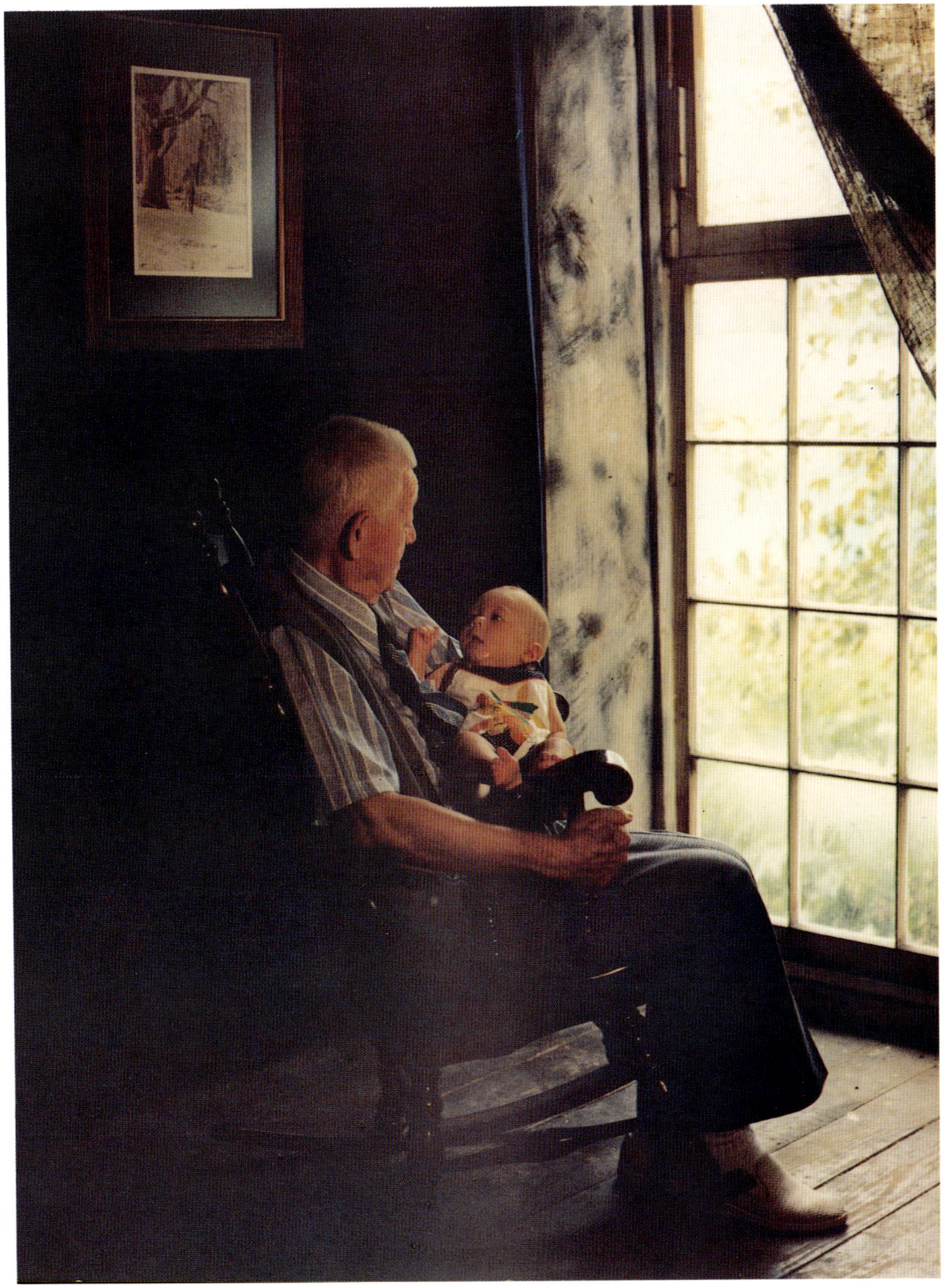

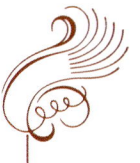

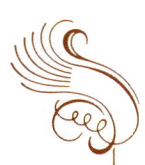

A First Date

The theater of a local community college was the setting of this portrait. Stage lighting highlighted Jim Warner's granddaughter and her friend playing the main characters on a first date.

15

First Day of School

A slightly different version of this portrait appears in *The Darker Brother.* Within the context of that book, which was published in 1974, the portrait presented a strong case against the isolation that was fostered by school segregation in the early 1970s.

In *Milestones,* that sense of isolation, suggested by the bright light and the shadow on the red brick wall, has less to do with color and more to do with a shy six-year-old's apprehension on the first day of school.

17

A Quiet Moment

Friends come to the studio to assist with a portrait for *Milestones* that will show a gentle, quiet moment shared between the "folks at home."

Papa rocks in his favorite chair, relaxed as he glances up from his paper. Mom is caught between words in their conversation as she cleans the lens of her eyeglasses.

The soft afternoon light of an overcast day was the only source of light for this portrait.

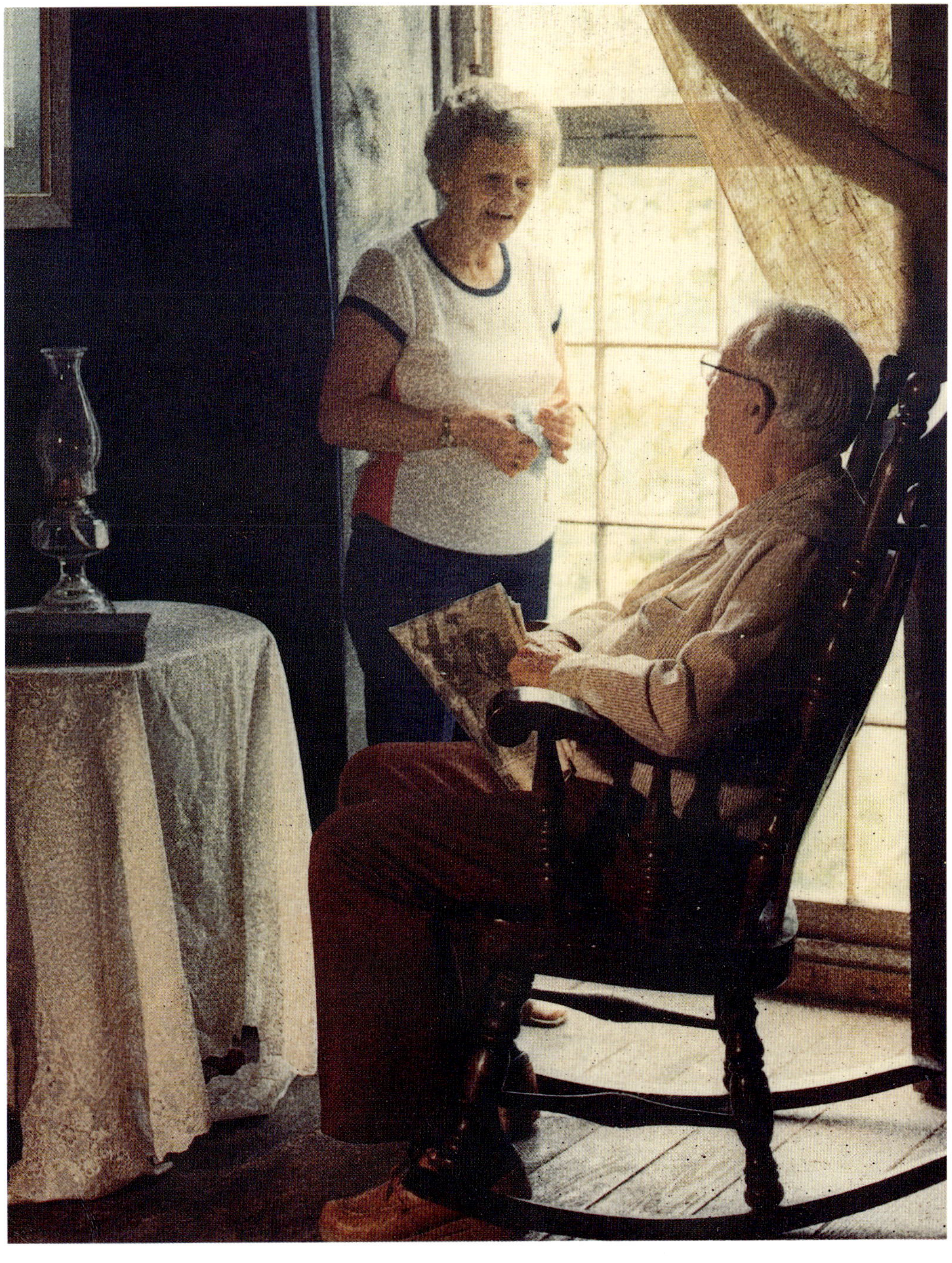

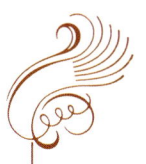

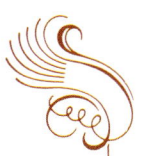

Anniversary Dinner

The owner of a local restaurant specializing in Spanish cuisine gave the photographers access to the restaurant's main dining room for this anniversary portrait. The couple sharing the moment were celebrating their 40th wedding anniversary.

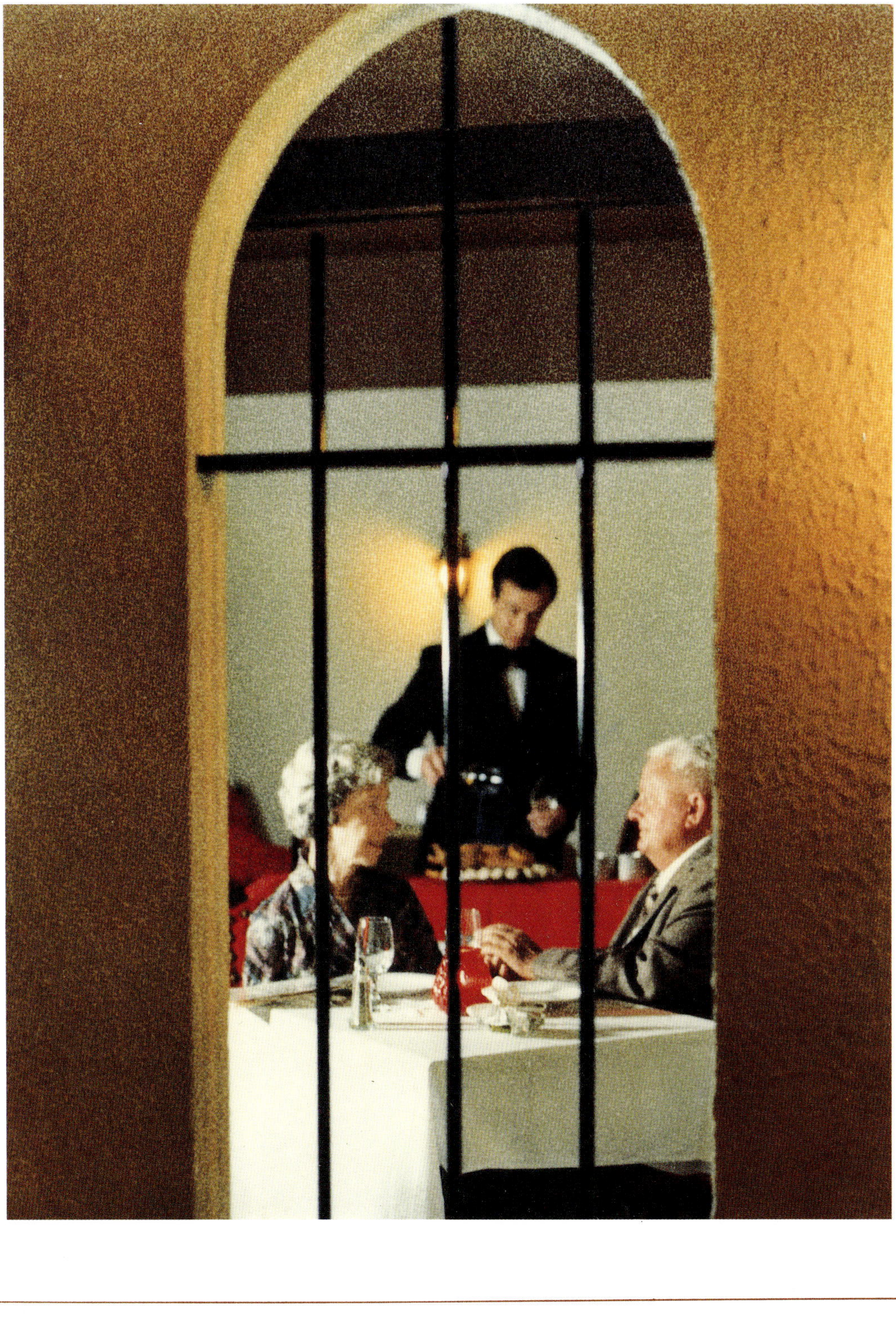

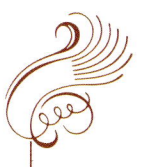

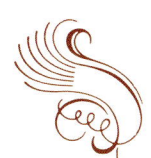

Waiting for the Phone to Ring

Here the photographers skillfully identify and portray a more subtle rite in our culture, the rite of communication. This portrait is from a studio collection titled "Means of Communication."

23

A Corner in the Attic

A corner of the studio became a corner in the attic for this popular gallery print, which was used in the photographers' inspirational book, *The Journey.*

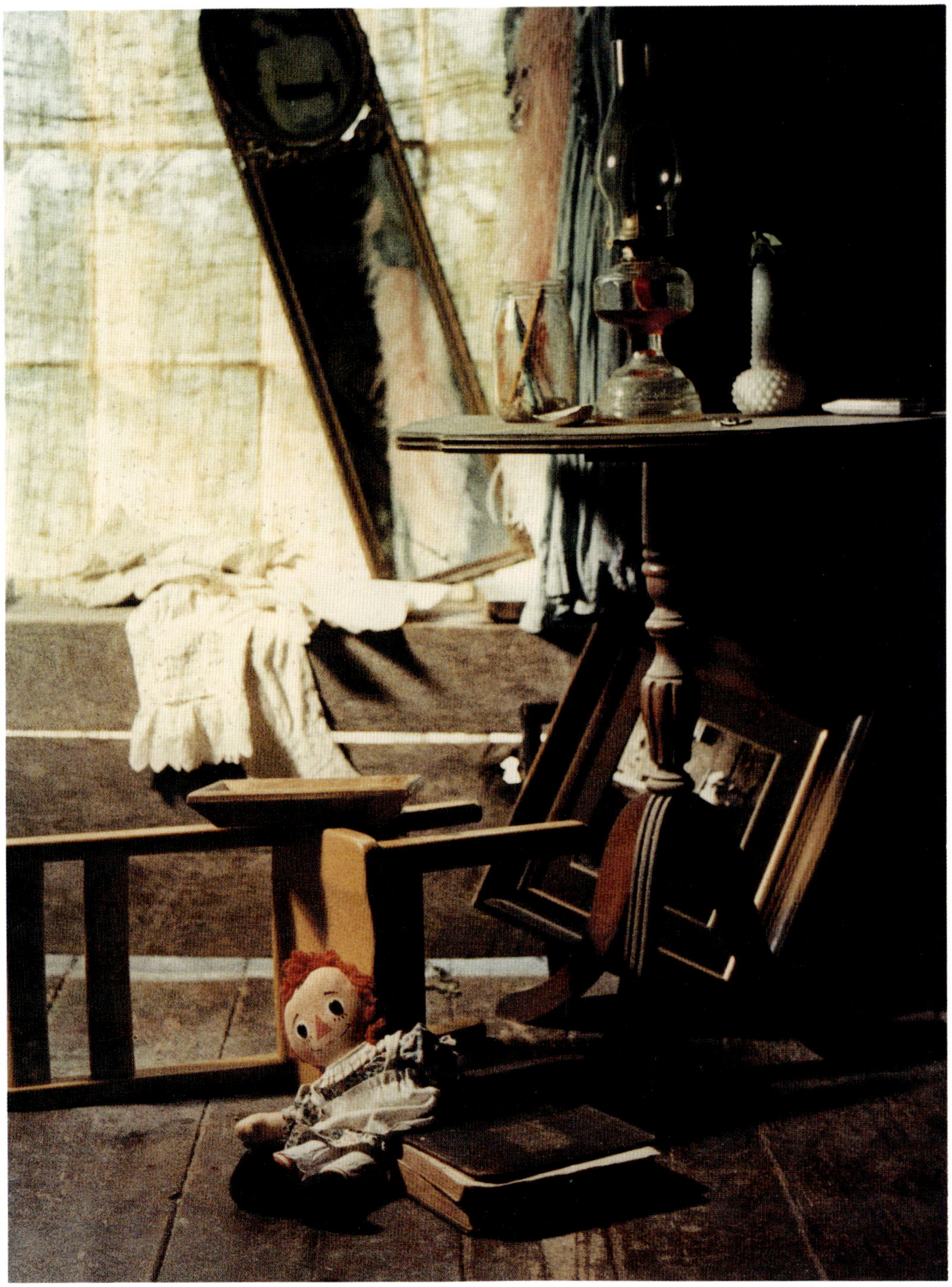

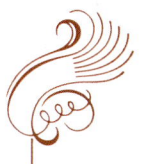

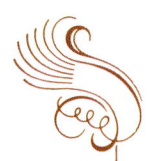

Read It Again, Grandpa

It was Christmas at the studio and the photographers' favorite Kris Kringle was painting visions of sugarplums for two little friends. By the end of the sitting, created for *Milestones,* the girls were calling old Kris "Grandpa Kringle."

27

Best Friends

Most of the time we choose our own friends. Sometimes, they choose us.

These are two friends who have been together a good many years. They posed for the photographers on a January day, shortly after an ice storm had glazed the countryside.

The portrait is from the *Best Friends* collection.

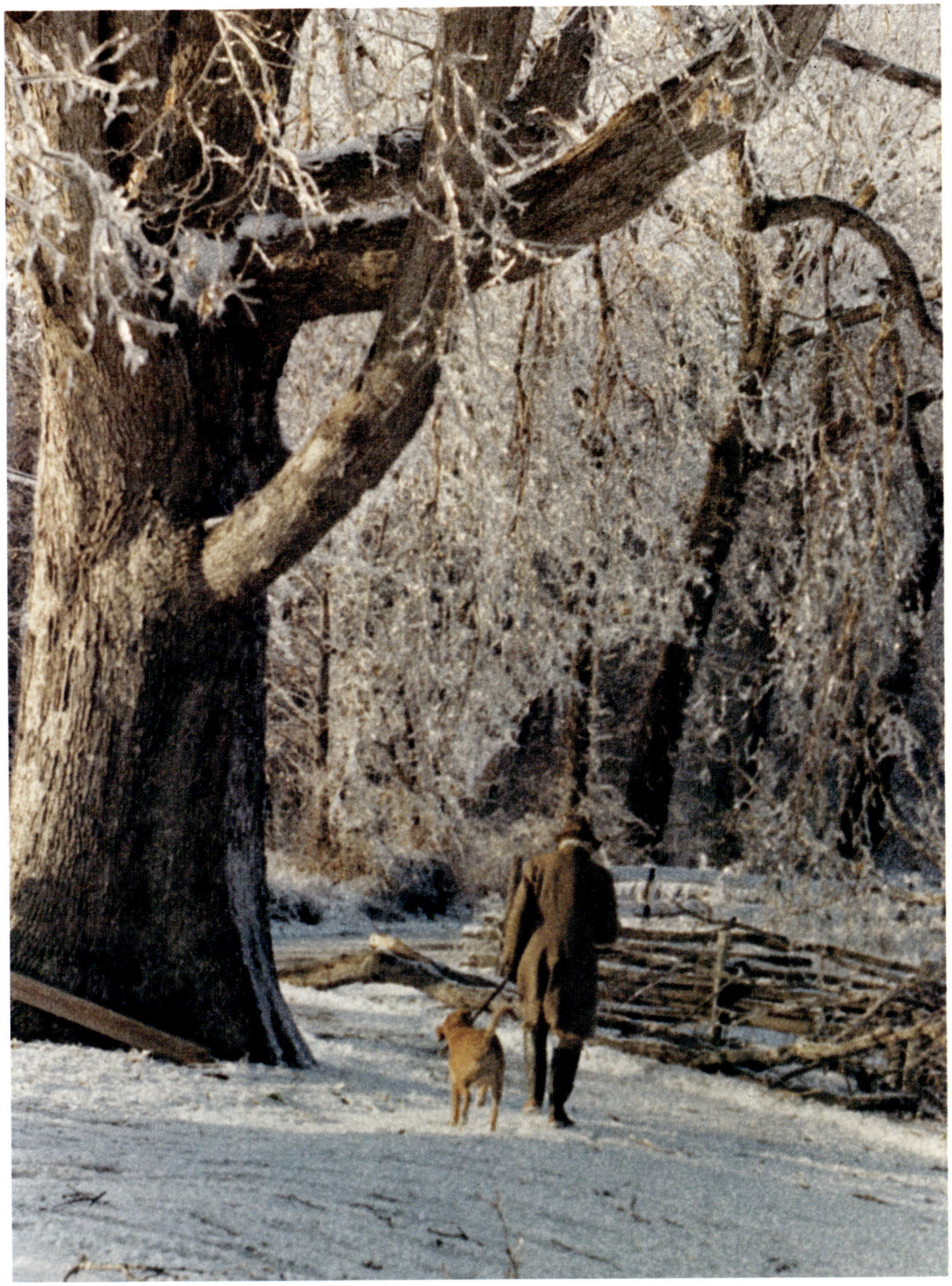

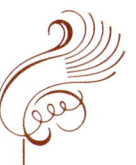

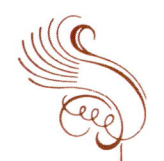

Act of Mercy

The setting is the office of the local veterinarian. The portrait is about comfort, companionship and kindness in old age.
"Act of Mercy" is from the *Best Friends* collection.

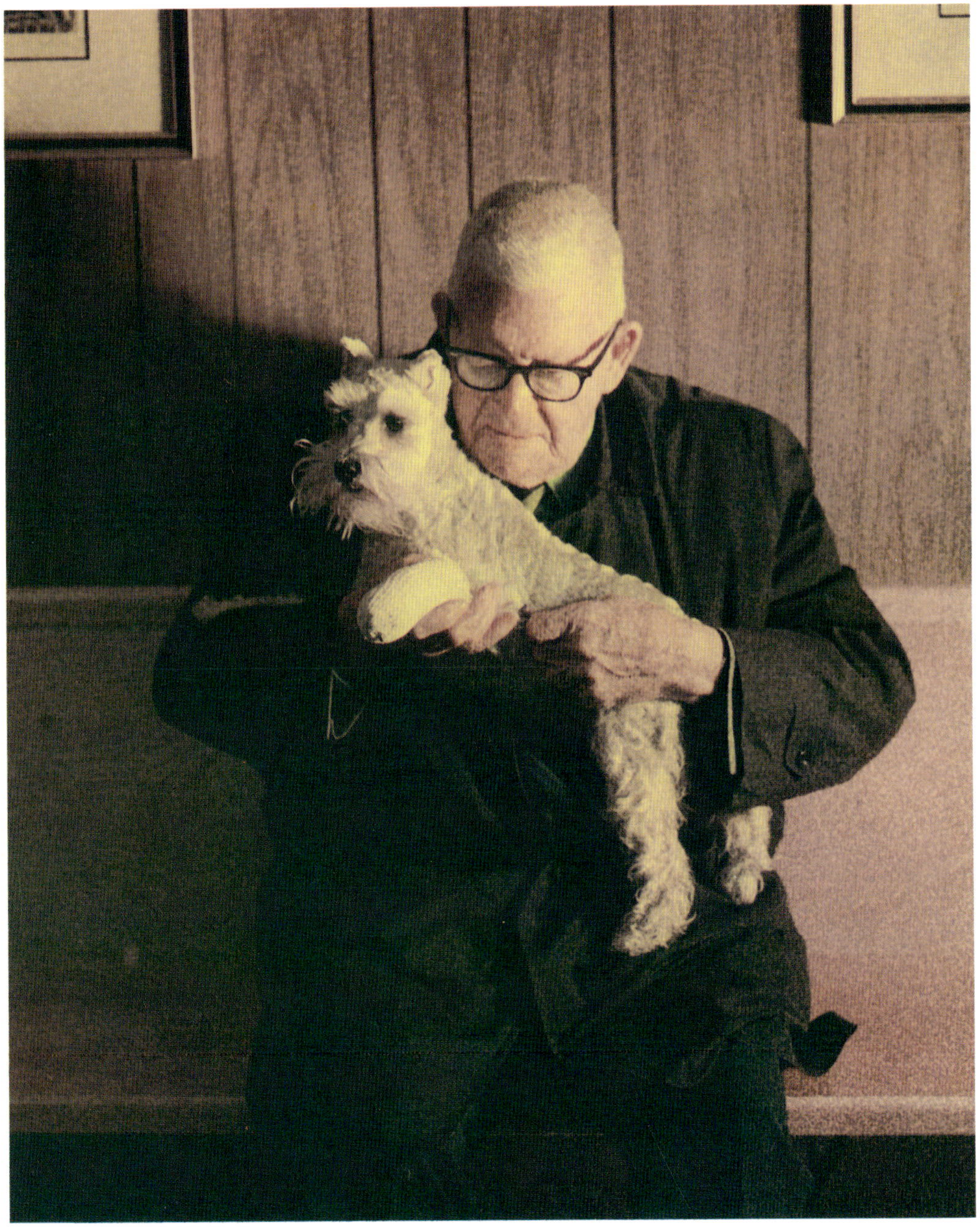

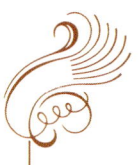

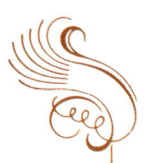

First Baby

A first baby usually demands and receives our utmost attention. This nursery scene is the moment of family introductions.

The nurse and the baby—an old and a new friend of the photographers—posed for the scene in a studio set. The portrait was made for *Milestones*.

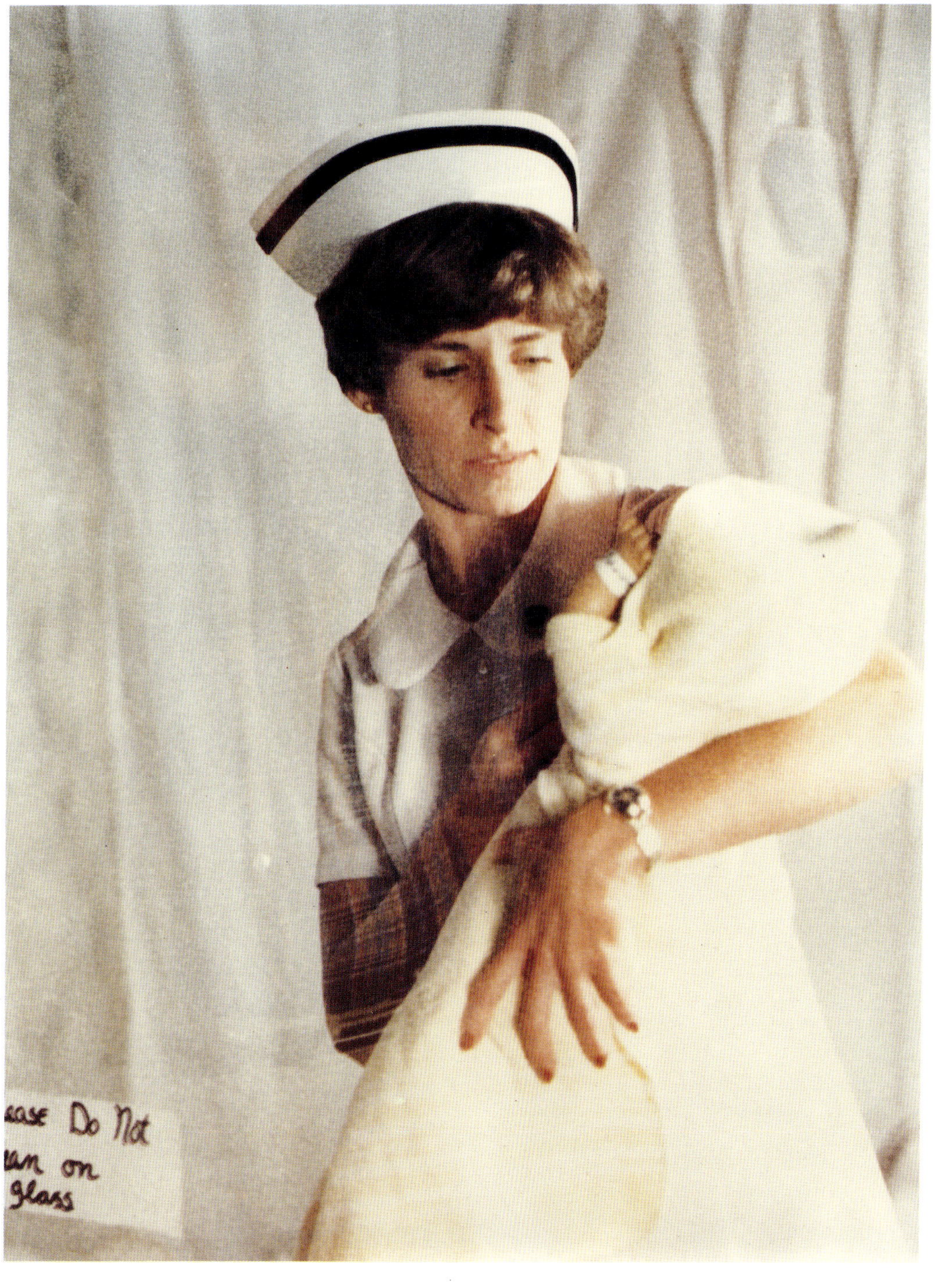
Do Not
on
Glass

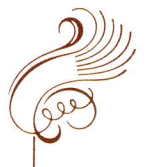

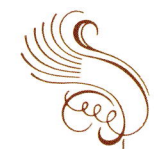

Two O'Clock Feeding

Memories of sharing this early-morning ritual with his wife when their children were infants gave Jim Warner the idea for this portrait.

The effect of the fog is appropriate for the memory of that hour of the morning.

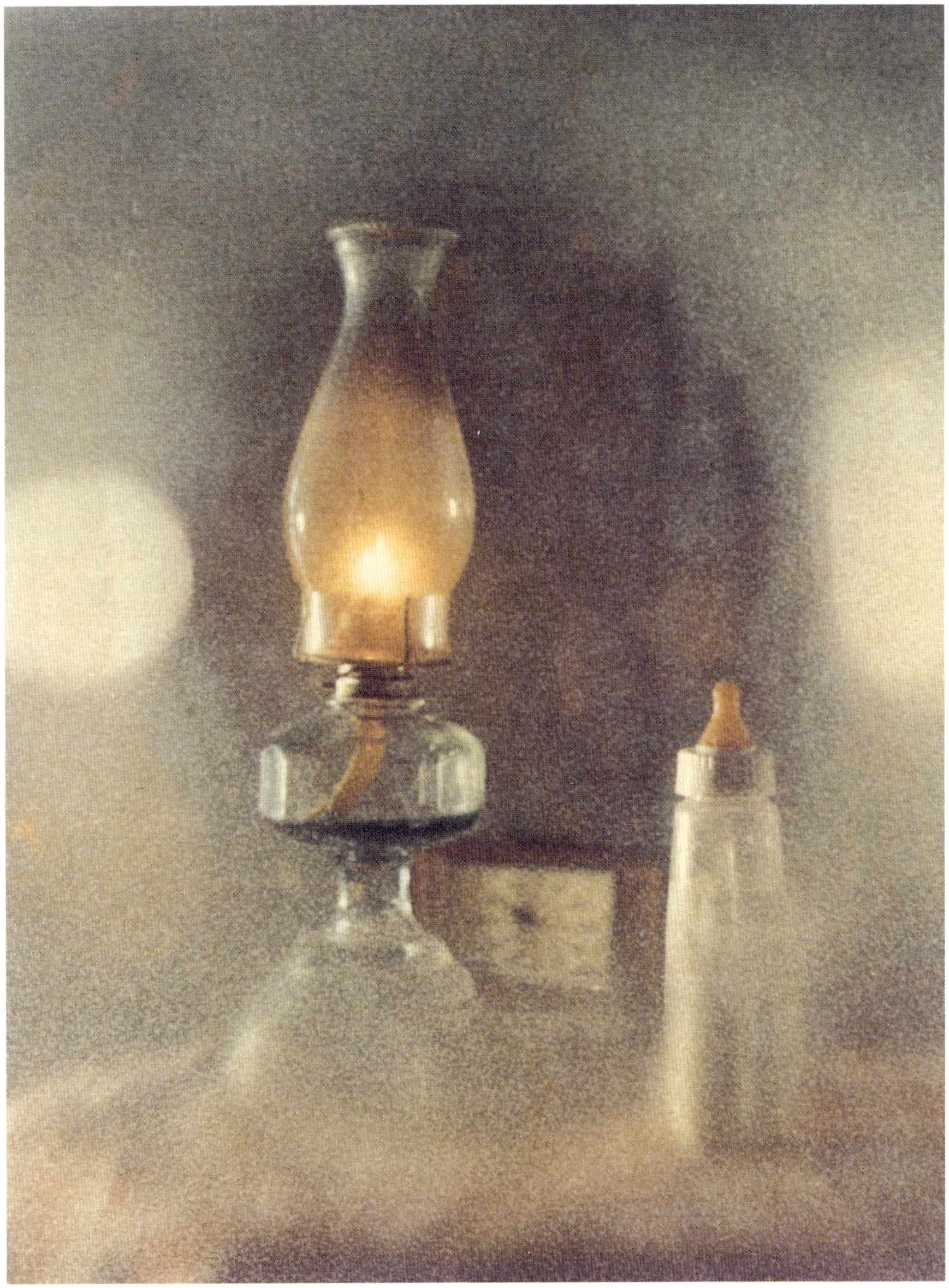

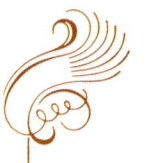

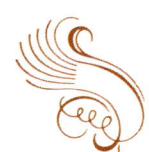

First Communion

A traditional rite of passage in the Roman Catholic Church, the first communion is an introduction into an active and participatory role in the liturgy of the mass.

For this portrait, the daughter of a friend and neighbor is wearing Margaret White's first communion dress, which Margaret wore in 1953.

37

Bar Mitzvah

This rite celebrates a young man's coming-of-age through the faith and traditions of his forefathers.

The portrait of this young man was taken at the studio as a remembrance of this special occasion.

39

Altar Boy

In the Catholic Church, it is an honor and a privilege to be chosen to serve as an altar boy.

This portrait is from a collection of prints titled "The Catholic Mass."

41

Scouting

A ceremonial rite of passage into ranking within the Girl Scouts is re-created at the studio with the help of two sisters involved in the scouting program. The younger sister, in the Brownie uniform, is caught in the reflection of a mirror as the ceremony for presenting an achievement award is rehearsed.

43

The Report Card

Body language tells the story of this day in the life of a student: anxiety on the part of the girl, concern on the part of the parent.
The scene is a studio set. Jim Warner's granddaughter and Margaret White are the characters.

45

Graduation Memories

A window seat at the studio becomes a corner for reflection. The alma mater, the varsity letter, the yearbook, and the graduate.

47

Goin' Fishin'

    This is a popular print from the studio's *Chesapeake* collection. The young boy heading down the road with fishing pole in hand is Jim Warner's grandson; the old gentleman out for a stroll is a neighbor and friend.
    Country roads are still a familiar sight in rural Maryland, where much of the studio's work is done. City lights and urban life are rare in the photographers' work.

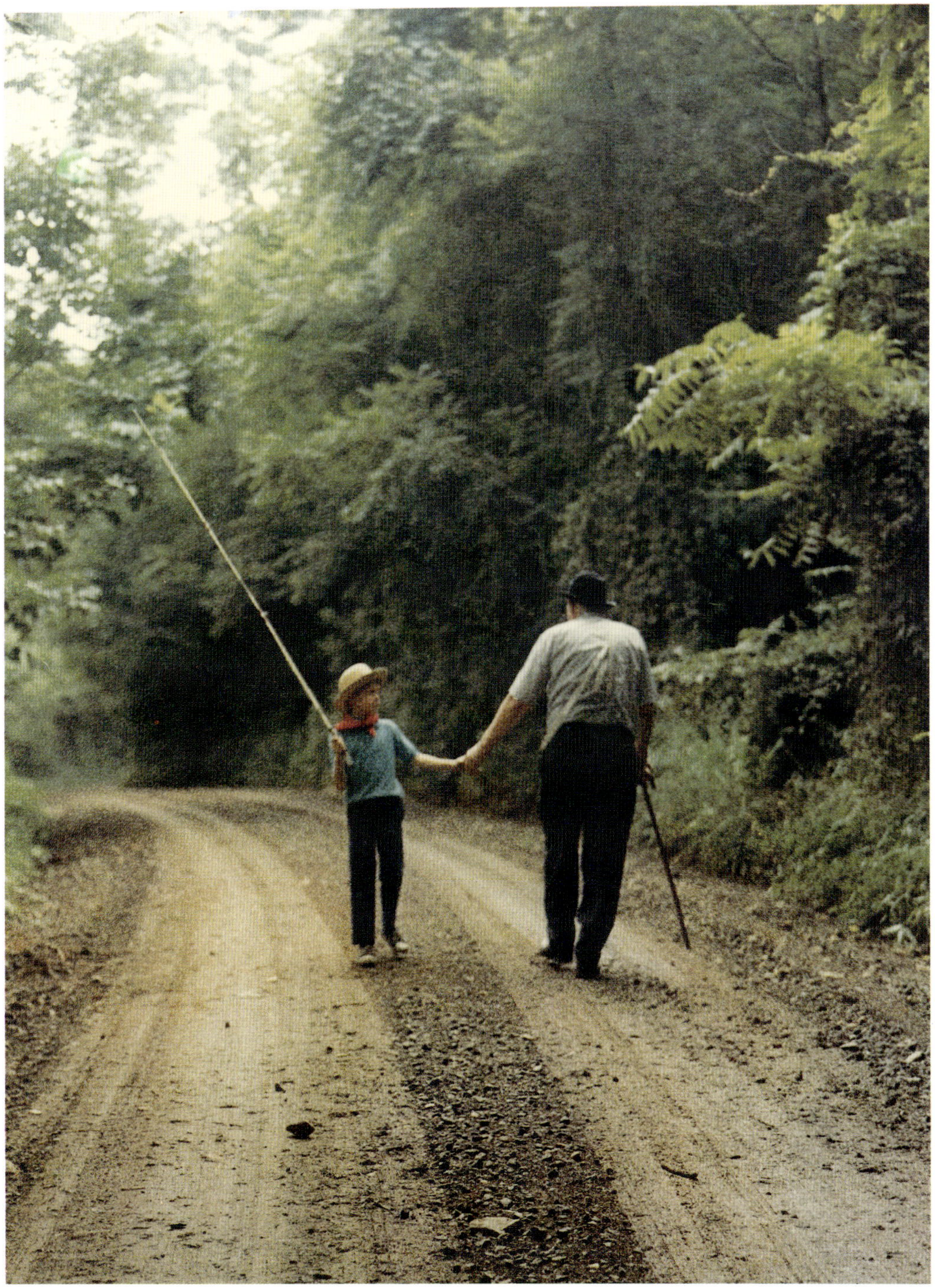

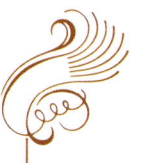

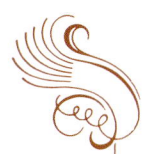

Let Me Teach You How to Pray

 Elements illustrating the Mormon belief in adults' responsibility for teaching the young and patriotism were incorporated into this scene to create an appealing portrait.
 The portrait was taken for the book jacket of *The Mormon Way*. However, since the publisher felt strongly about using only portraits of Mormons in the book and neither the girl, Jim Warner's granddaughter, nor the friend posing as her teacher were Mormon, the print ended up on the editor's floor.

51

I Learned It at My Mother's Knee

Nature provided the setting for this portrait: a sunlit meadow visible through a hedgerow just beyond the studio doorway. The idea behind the portrait—to capture a gentle, quiet moment between a mother and daughter—lent itself to the feel of a soft-focus print. The leaves of the hedgerow supplied the soft outer edges.

"I Learned It at My Mother's Knee" was done for the book *The Journey*.

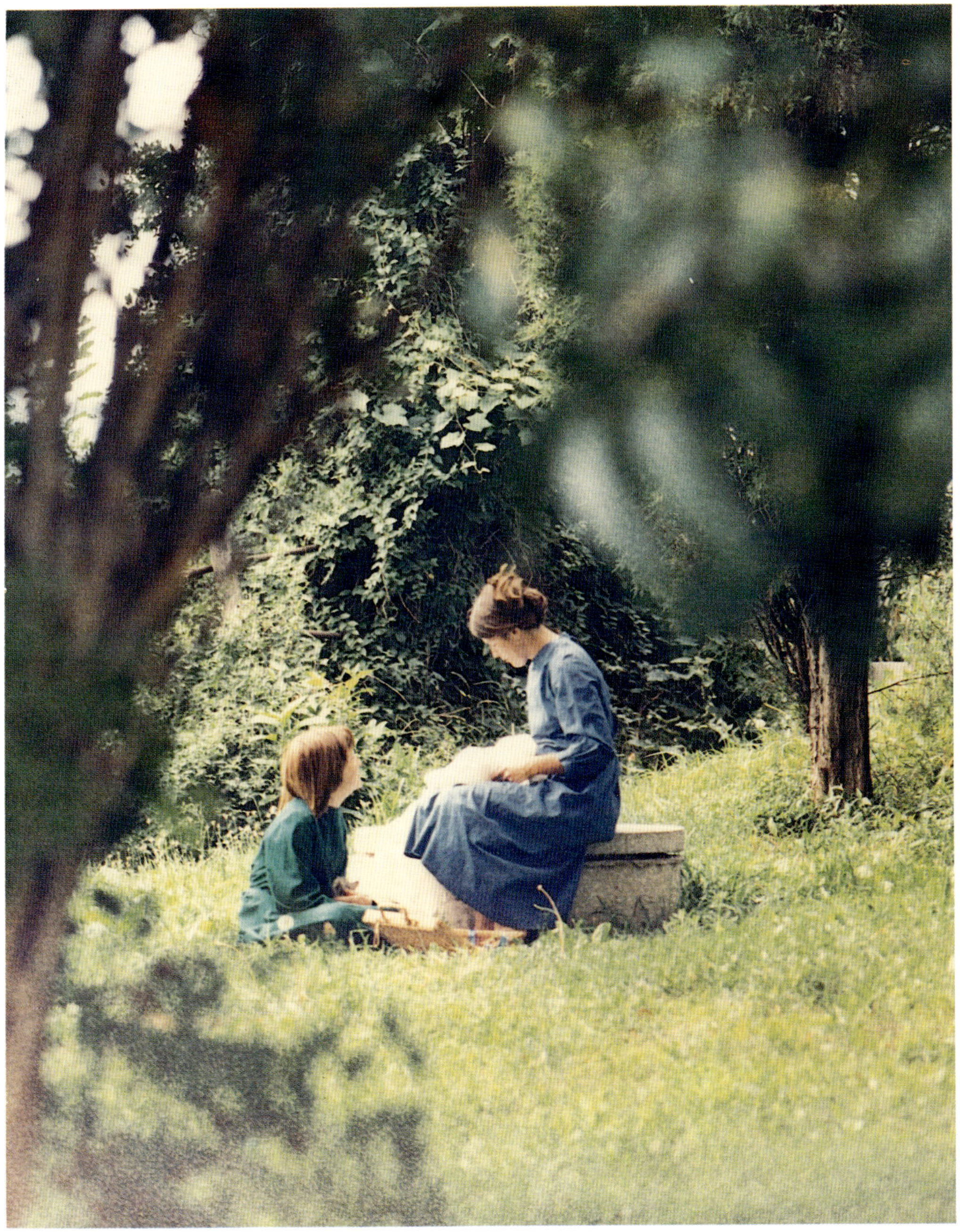

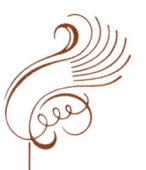

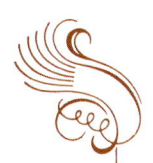

Daydreaming

An unguarded moment created a striking pose for the book *The Darker Brother.* The strength of the portrait was matched by text from "The Declaration of Rights of the Child":

He shall be brought up in a spirit of understanding, tolerance, friendship among peoples…and in full consciousness that his energy and talents should be devoted to the service of his fellow man.

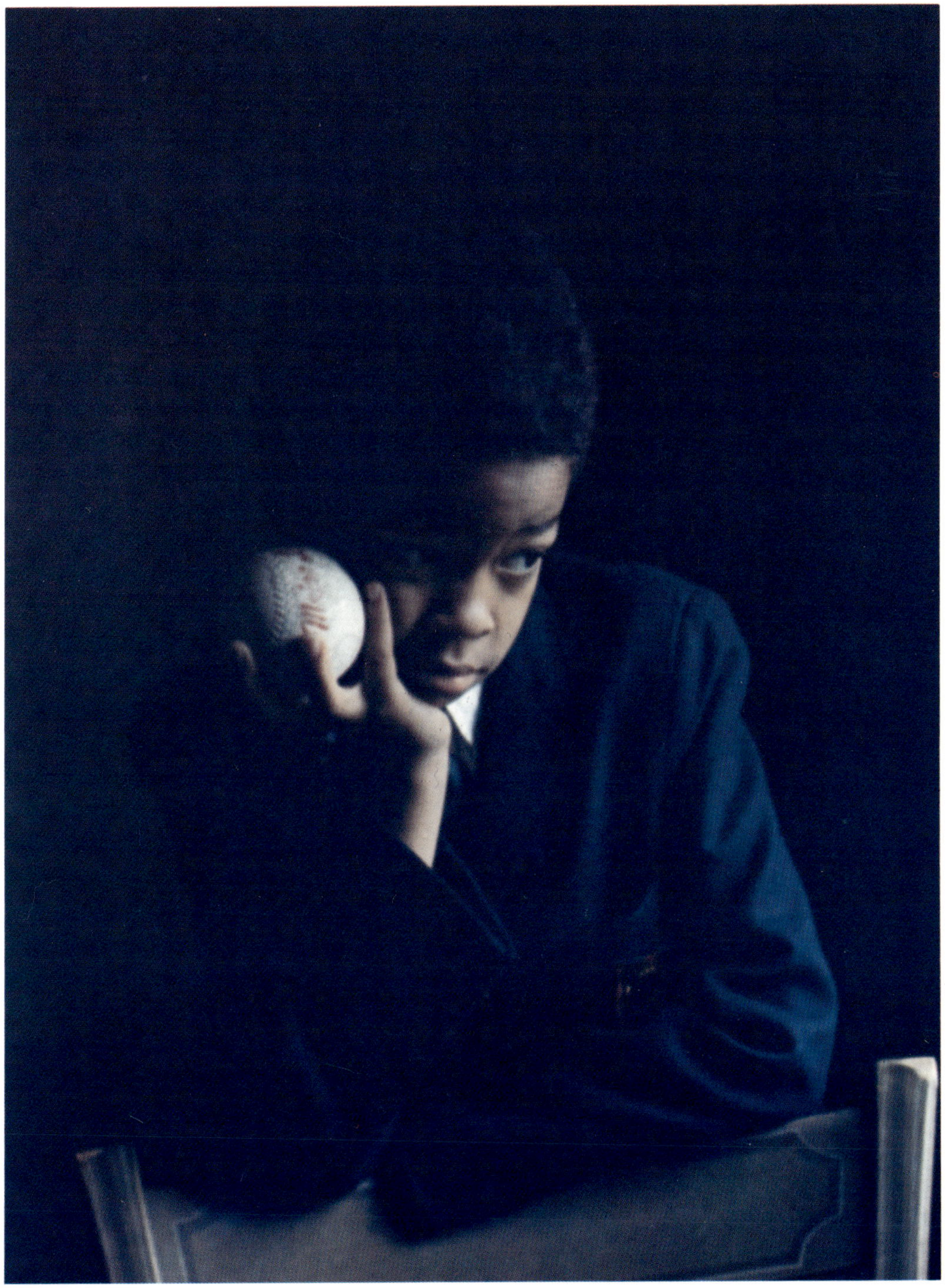

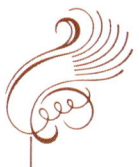

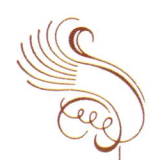

Running Away from Home

Another unguarded moment in the life of a child. This one cannot be recorded but can only be preserved in memory.

Here, the memory is translated into a portrait that was created for the photographers' book *Best Friends*.

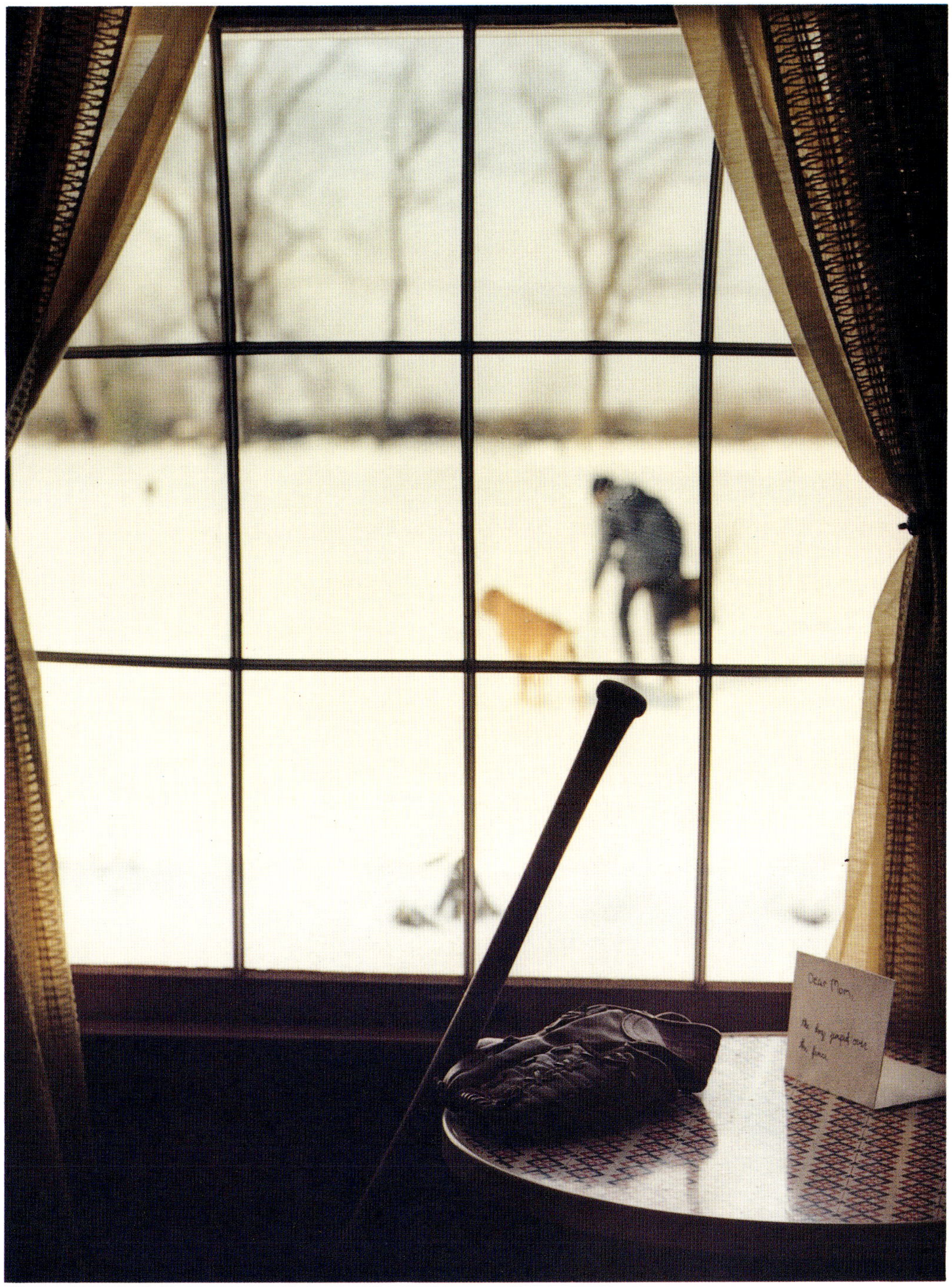
Dear Mom,

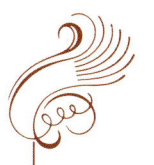

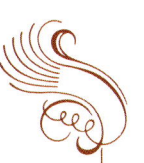

When I Was a Child

A thought-provoking portrait from the book *The Darker Brother*, this photographic interpretation of a poem tells the story of a young boy who experiences racial prejudice for the first time in his life.

The color of the boy's skin incites others to hurl abusive words at him. His bike is taken away, and he has to beg to get it back. He then comes home, throws his bicycle on the ground, and settles down to think through conflicting notions of what his future will hold.

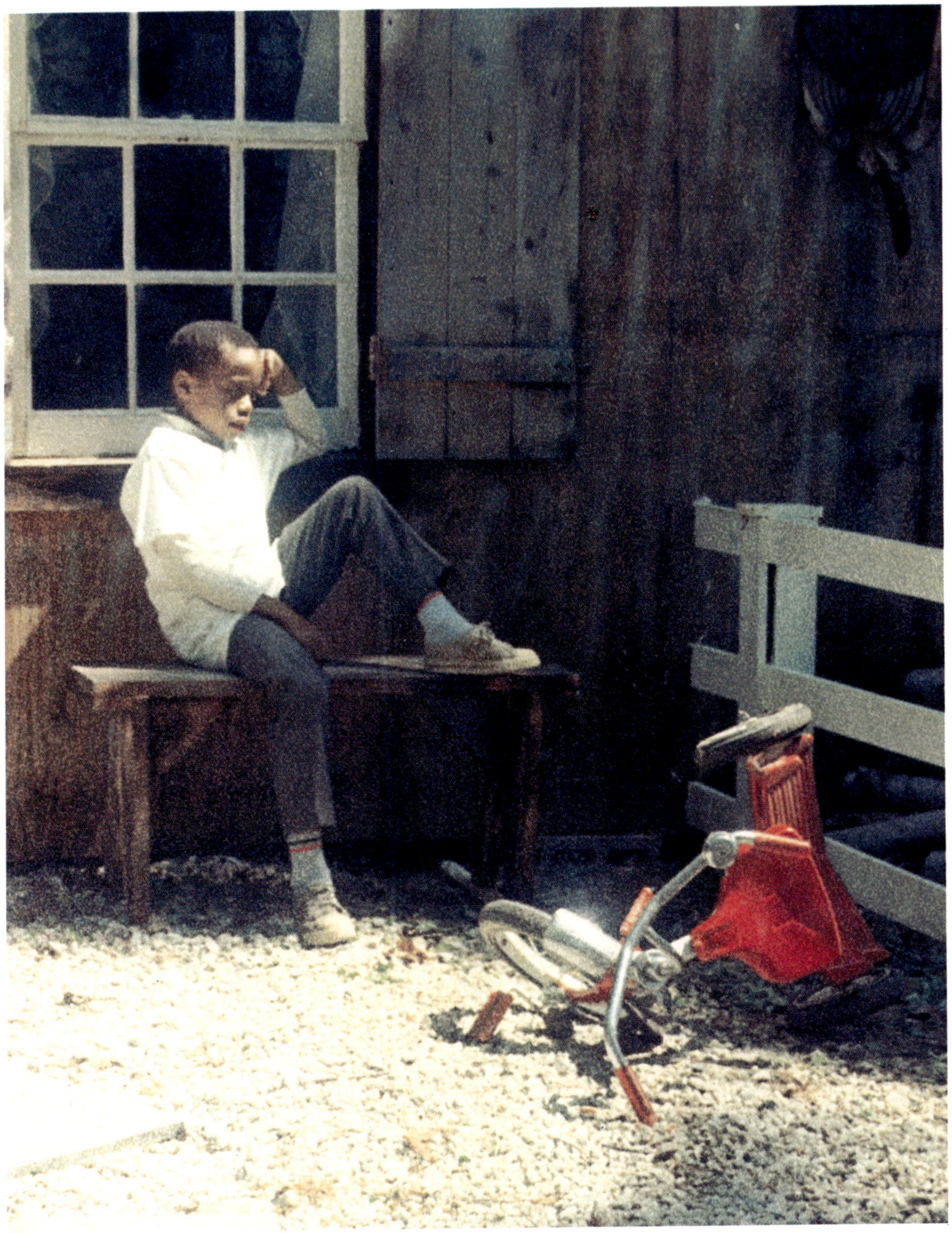

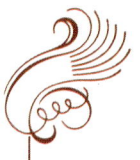

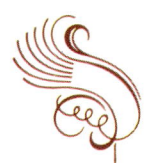

Dunkard Baptism

    In search of unusual rites, the photographers went to Lancaster, Pennsylvania, to witness a Dunkard baptism.
    The countryside of Lancaster County was familiar territory for Jim Warner; it was there that he had found the Amish community that became the subject of his first two books, *The Gentle People* and *The Quiet Land*.

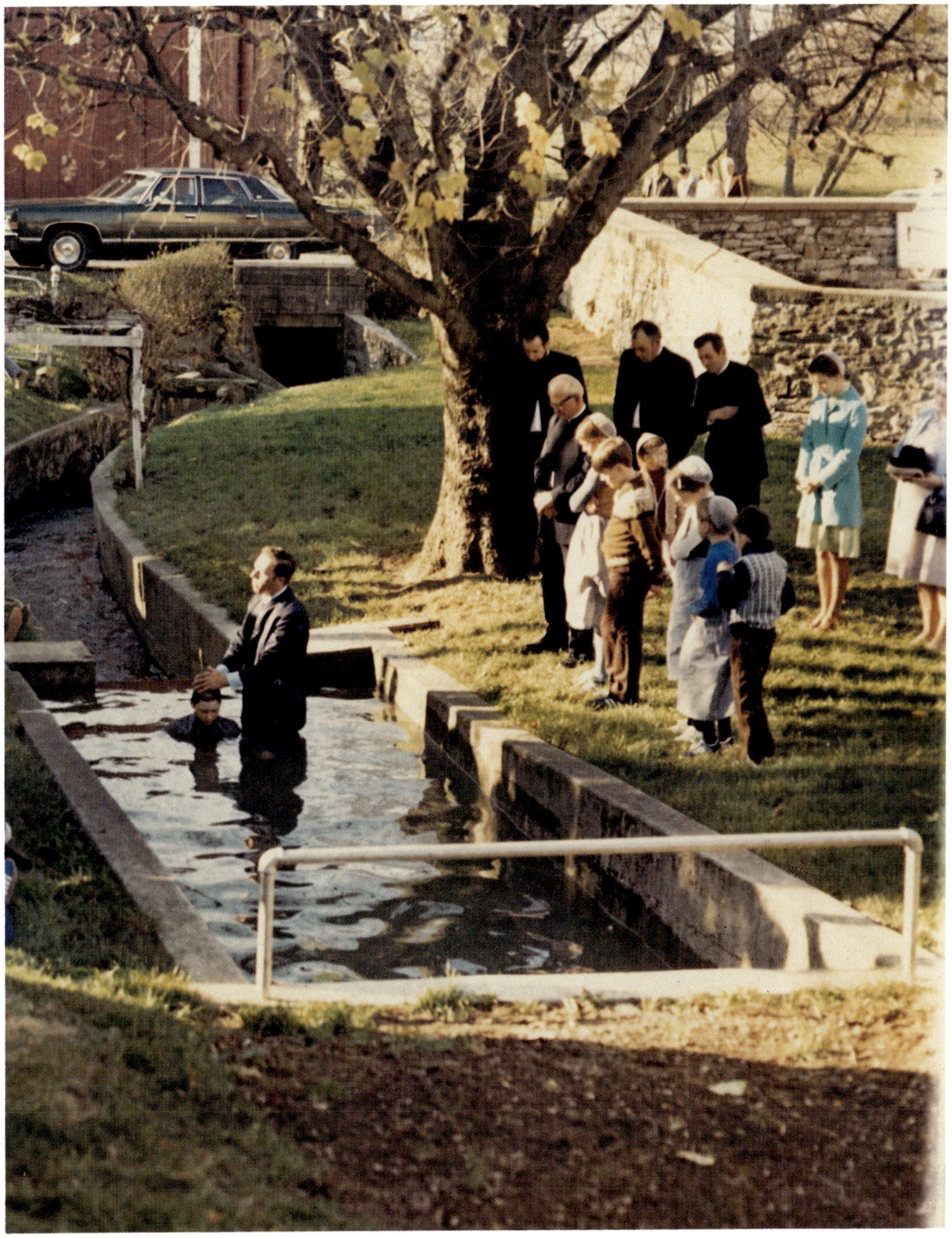

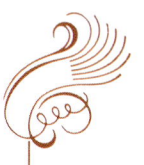

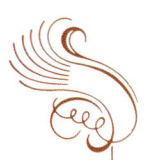

Japanese Wedding Ceremony

A Tenrikyo wedding ceremony was re-created for the photographers at the home of Japanese friends in nearby Washington, D.C.

Tenrikyo is the teaching of Tenri-Ō-no-Mikoto, or God the Parent, as believed by members of this Japanese religious sect.

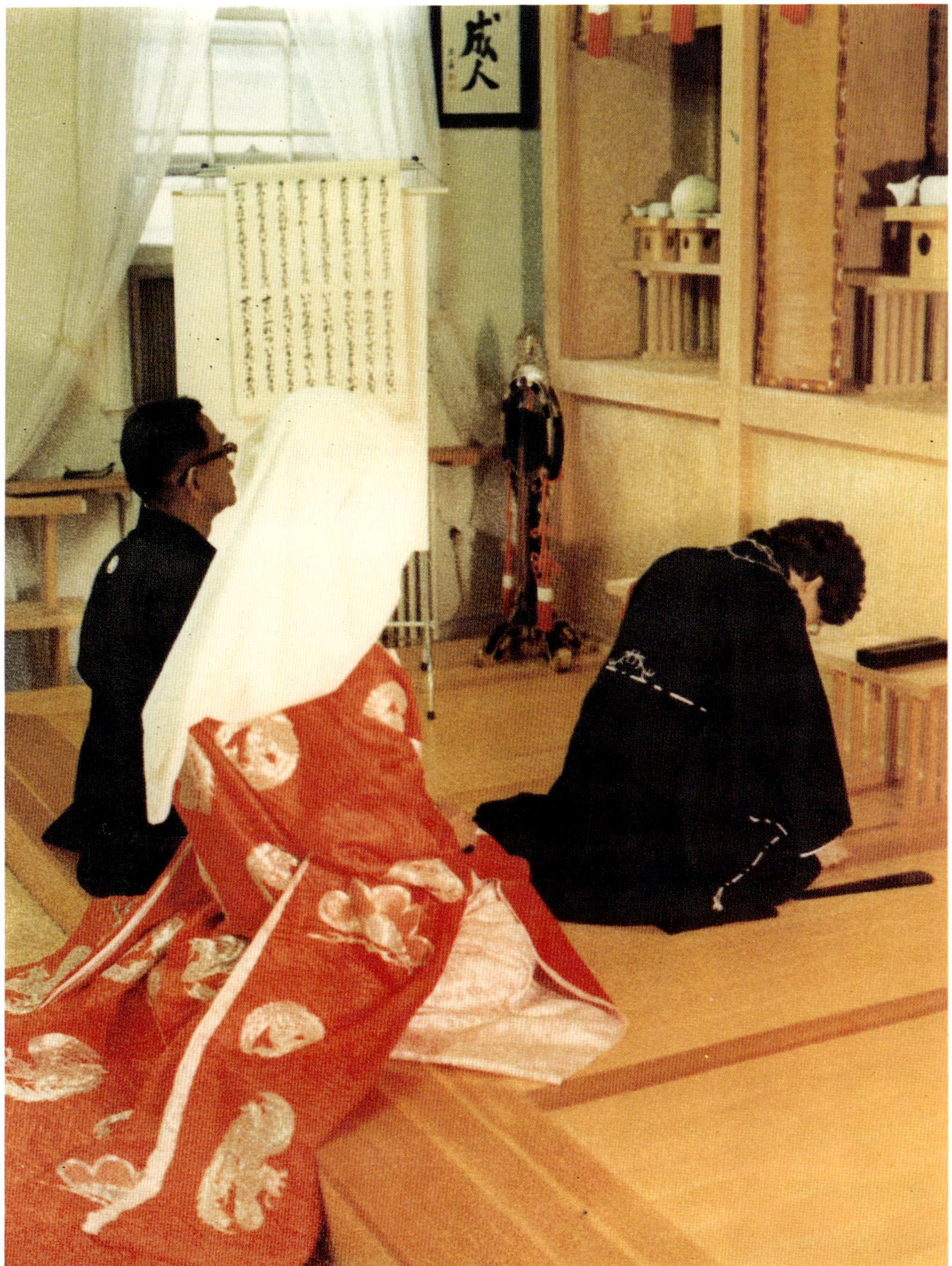
成人

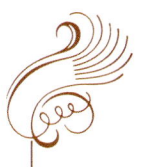

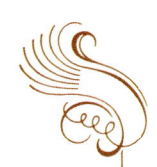

Queen of the Prom

A high school girl's dream-come-true, a memory never to be forgotten. A crown. A title. A bouquet of roses. One stem will be pressed between the pages of her diary and saved for the rest of her life.

65

Happy Hour

The after-work cocktail. The Happy Hour set. A popular 20th-century scene.

This portrait might well be called "One Photographer in the Light of the Other," since Jim Warner was behind the camera while Margaret White posed for the shot of this contemporary rite.

67

Veterans Day

    The old soldier and the young patriot enjoy the parade as the flag passes by.
    The scene, re-created at the studio doorway, is one from a small American town on a day when we remember soldiers, the nation's defenders, and youth, the nation's defenseless.

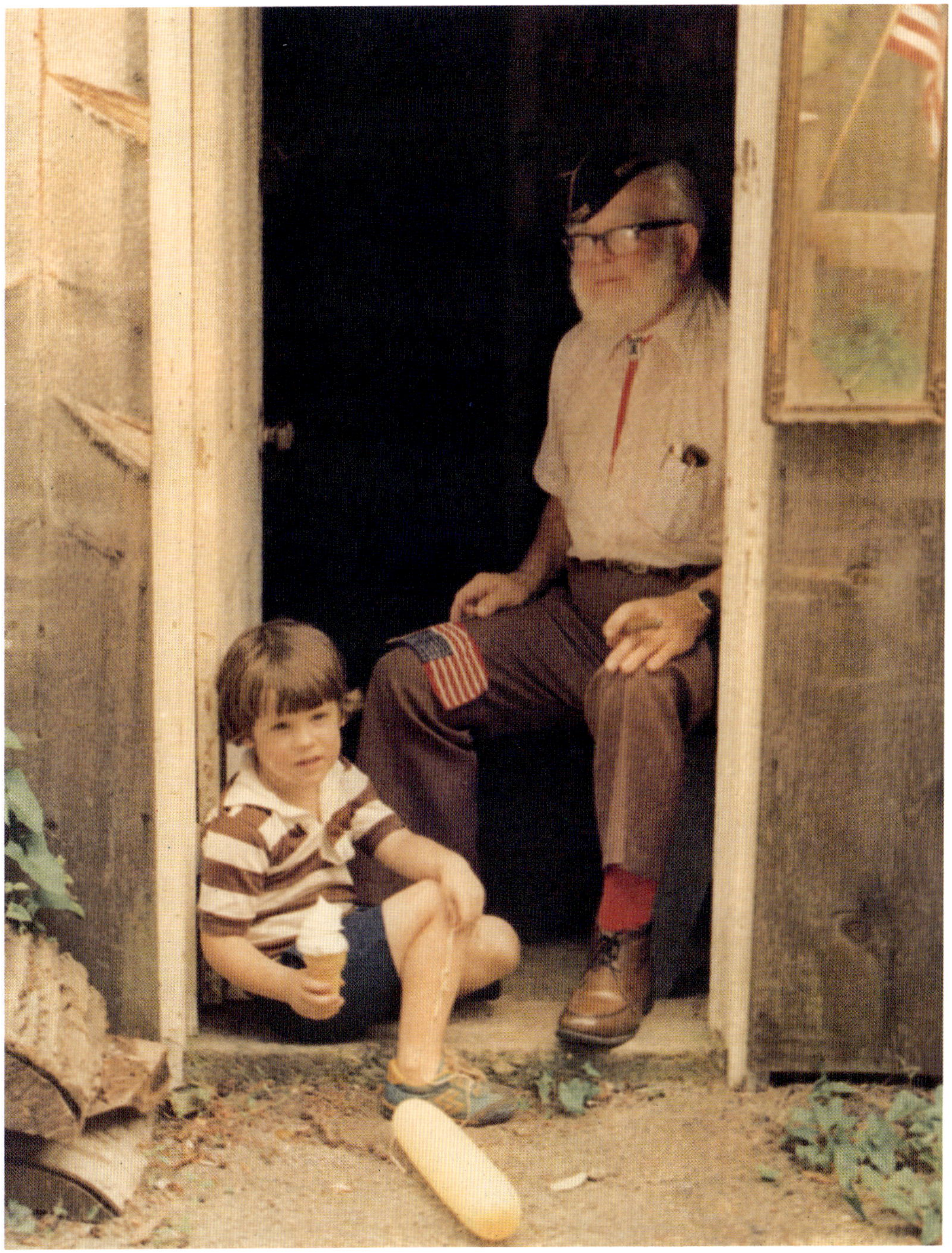

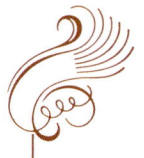

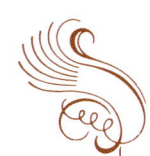

Lemonade Stand

The roadside lemonade stand is often a child's first business venture. Such a portrait, if taken at all, becomes part of the family album.

Once again, in a country setting.

71

Grand Opening

The shopkeeper, the restaurateur, the entrepreneur in pursuit of a dream. To be master of one's own ship, captain of one's own fate.

Here, last-minute details for the grand opening of a local business venture are tended to.

73

Carving Pumpkins

Fall has come to the country; dried cornstalks, October leaves, and pumpkin carving are all memories of the rites of autumn.

The youngster in this portrait is carving pumpkins on a set in the studio. Vignetting was used to create distance between the subject and the observer.

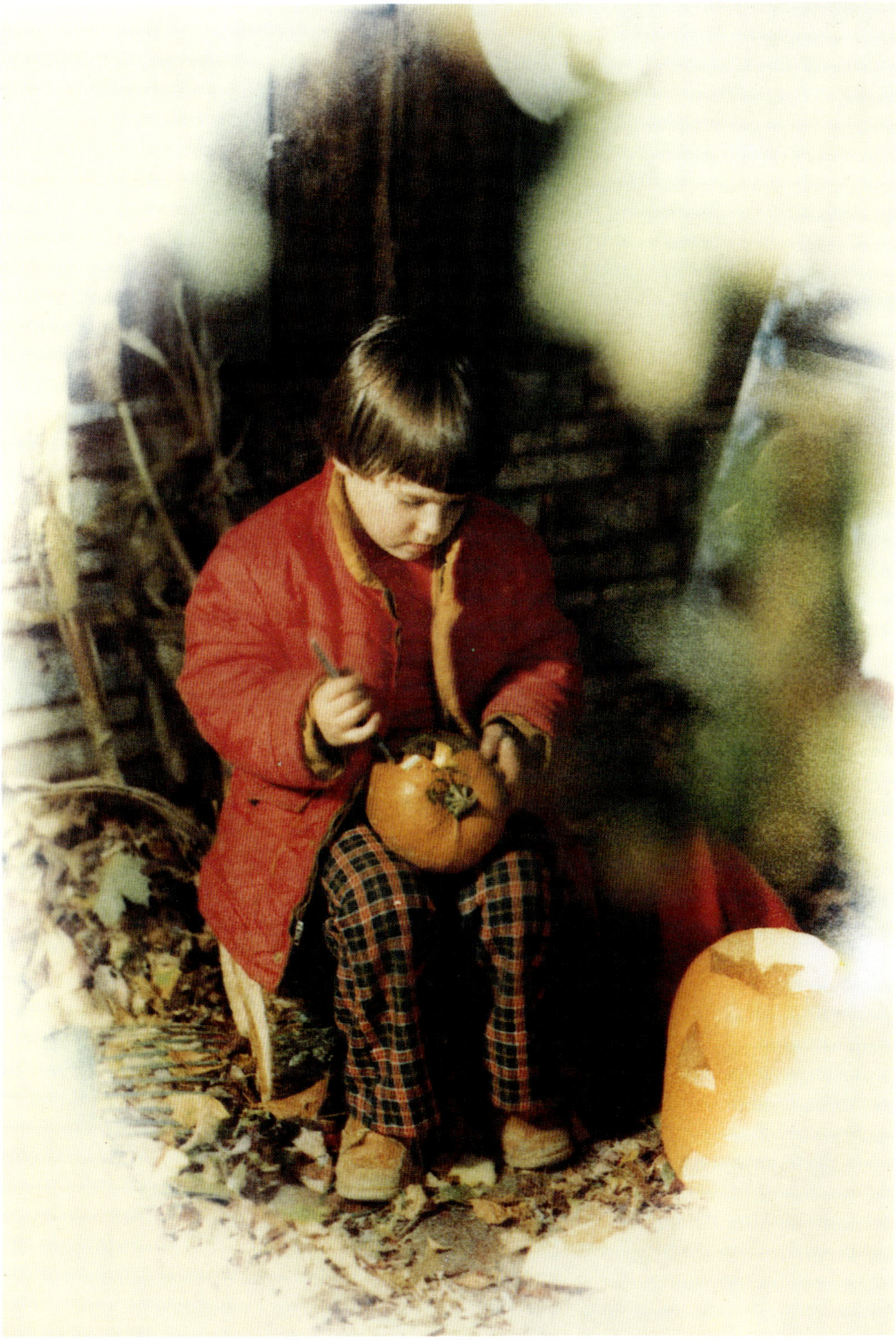

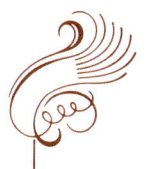

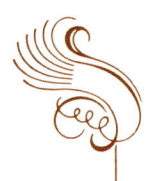

Trick or Treating

It's Halloween! Dracula comes to the door for a trick or a treat–and leaves with a portrait.

Costume and makeup were done by a theater friend and studio set designer, Gary Ensor. The young Dracula is Margaret White's son.

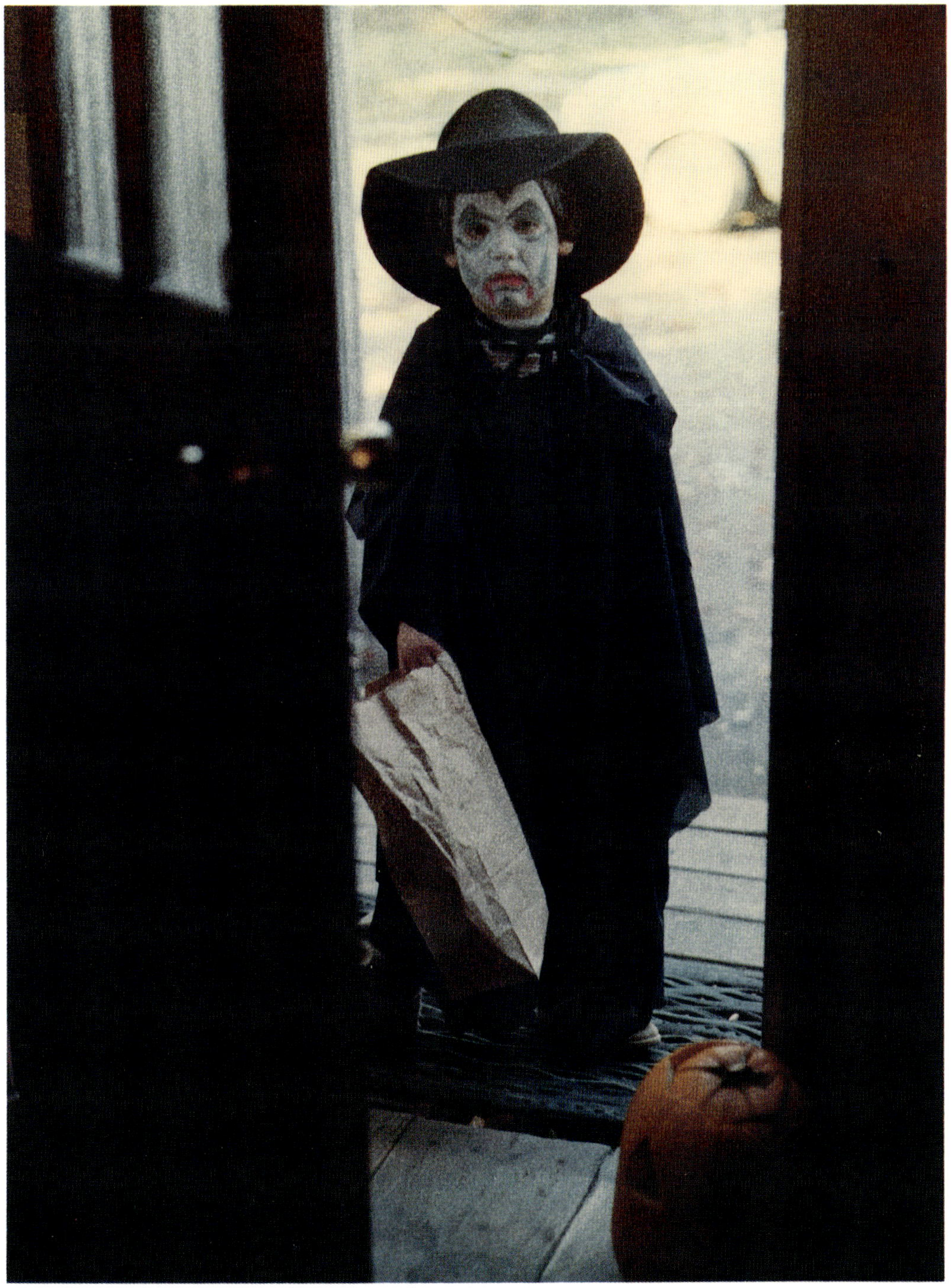

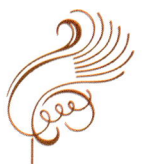

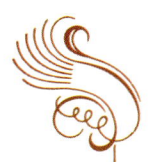

Salvation Army

    The sight of the Salvation Army bell ringers on street corners and in front of department store windows reminds us of the season of giving and the sounds of that season, carols, bells, and clinks of coins being tossed into the black iron pots of these caretakers of the homeless and poor.

    This is a scene slowly fading from society's view. Here, for *Milestones,* it is re-created as we remember it.

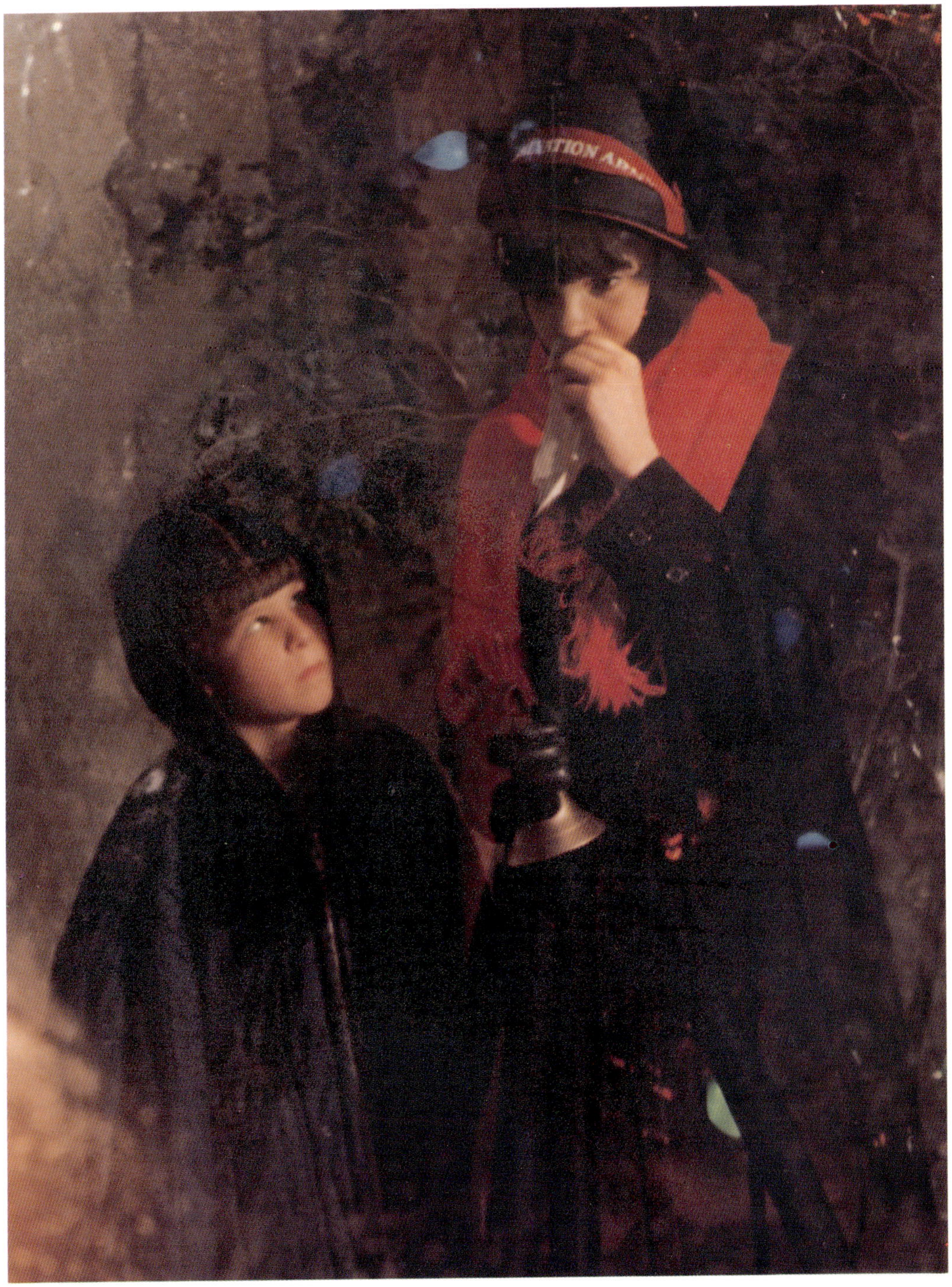

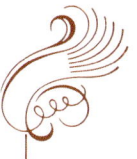

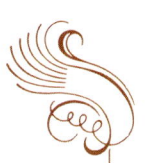

Jury Duty

One rite in our justice system is calling citizens to serve on jury duty.
A town courthouse is the setting for this scene.

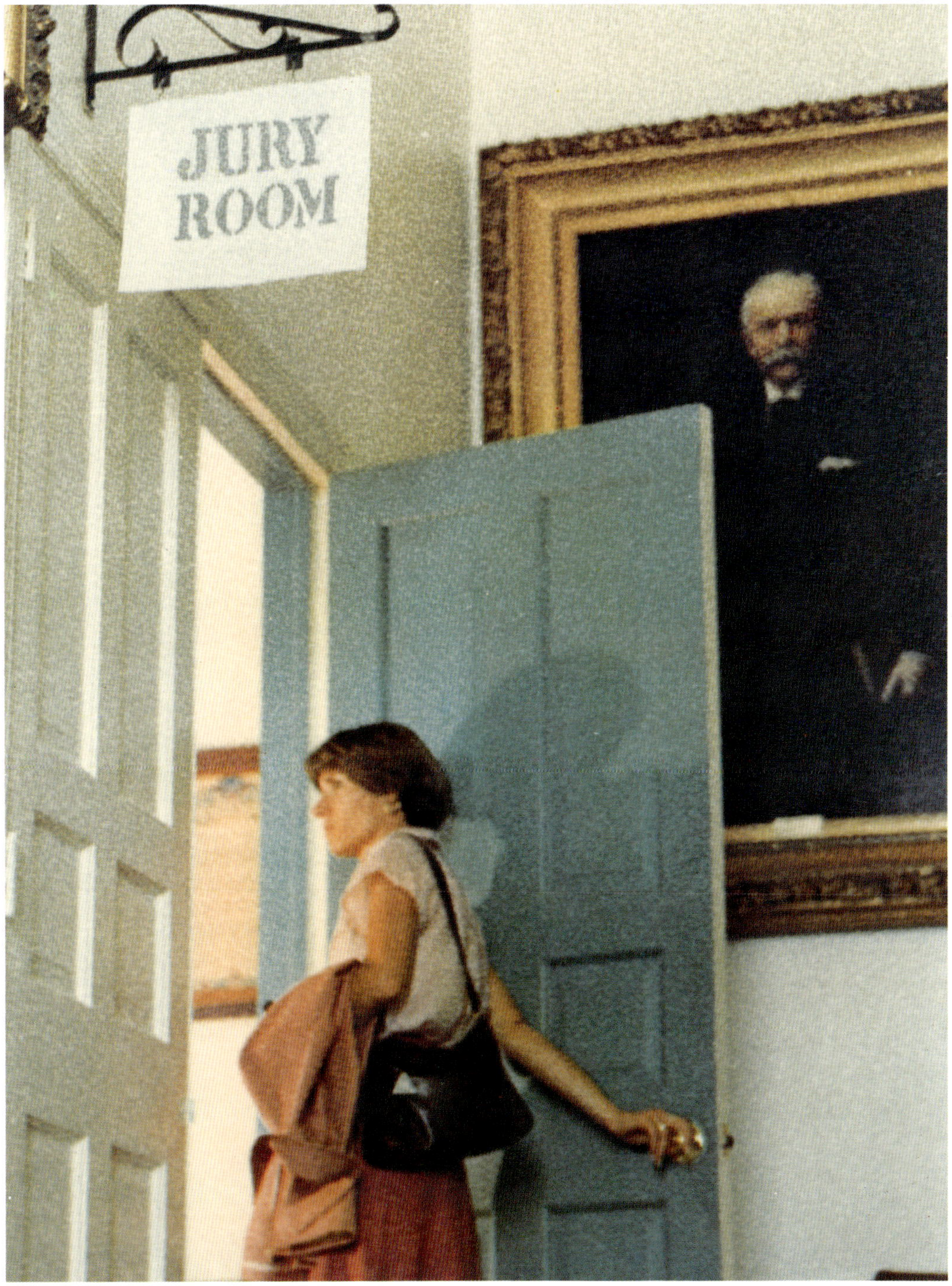
JURY
ROOM

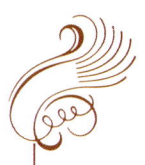

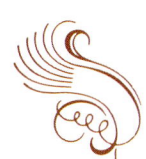

Introduction into the Service

The swearing-in ceremony is a citizen's rite of passage from civilian to military life. It is a man's or a woman's personal response to the call to serve in the Armed Forces.

The photographers chose a local Naval Recruit Center to capture this scene.

83

Voting

**At the bottom of all the tributes paid to democracy is the little man, with a little pencil, making a little cross on a little bit of paper—no amount of rhetoric or voluminous discussion can possibly diminish the overwhelming importance of the point.
Sir Winston Churchill, speech,
House of Commons, October 31, 1944**

The ballot box used in this portrait was retrieved from the basement of the local county courthouse. It had not been used for many years, since the modern voting machine came into use. However, there are still small towns across America where this scene would not be unfamiliar.

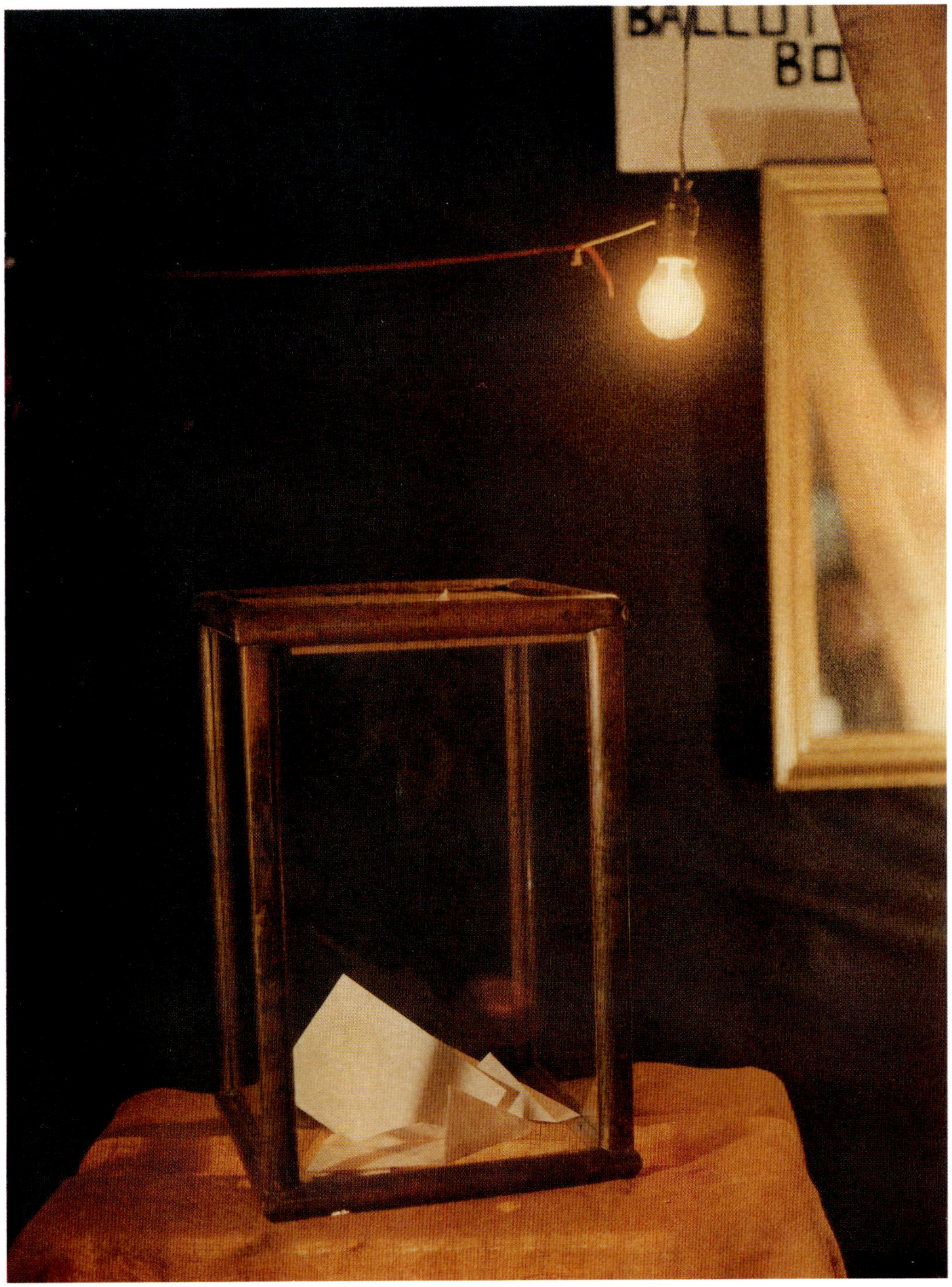

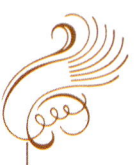

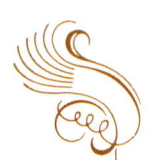

Running for Office

Memorabilia from the campaign trail. Reelection signs, campaign hats, and campaign buttons are stashed away among the collectibles that become part of our personal histories.

87

On Strike

Better pay, better working conditions, increased benefits, and stronger lines of communication between management and employees are among the demands of workers who band together to strike.

A march by local union workers captured the photographers' interest for *Milestones*.

89

Moments at Prayer: Praying the Rosary

Votive candles provide the only light for this dramatic study of a woman praying with her rosary beads. The play of flickering light on the woman's hands and face gives the portrait its drama and strength.

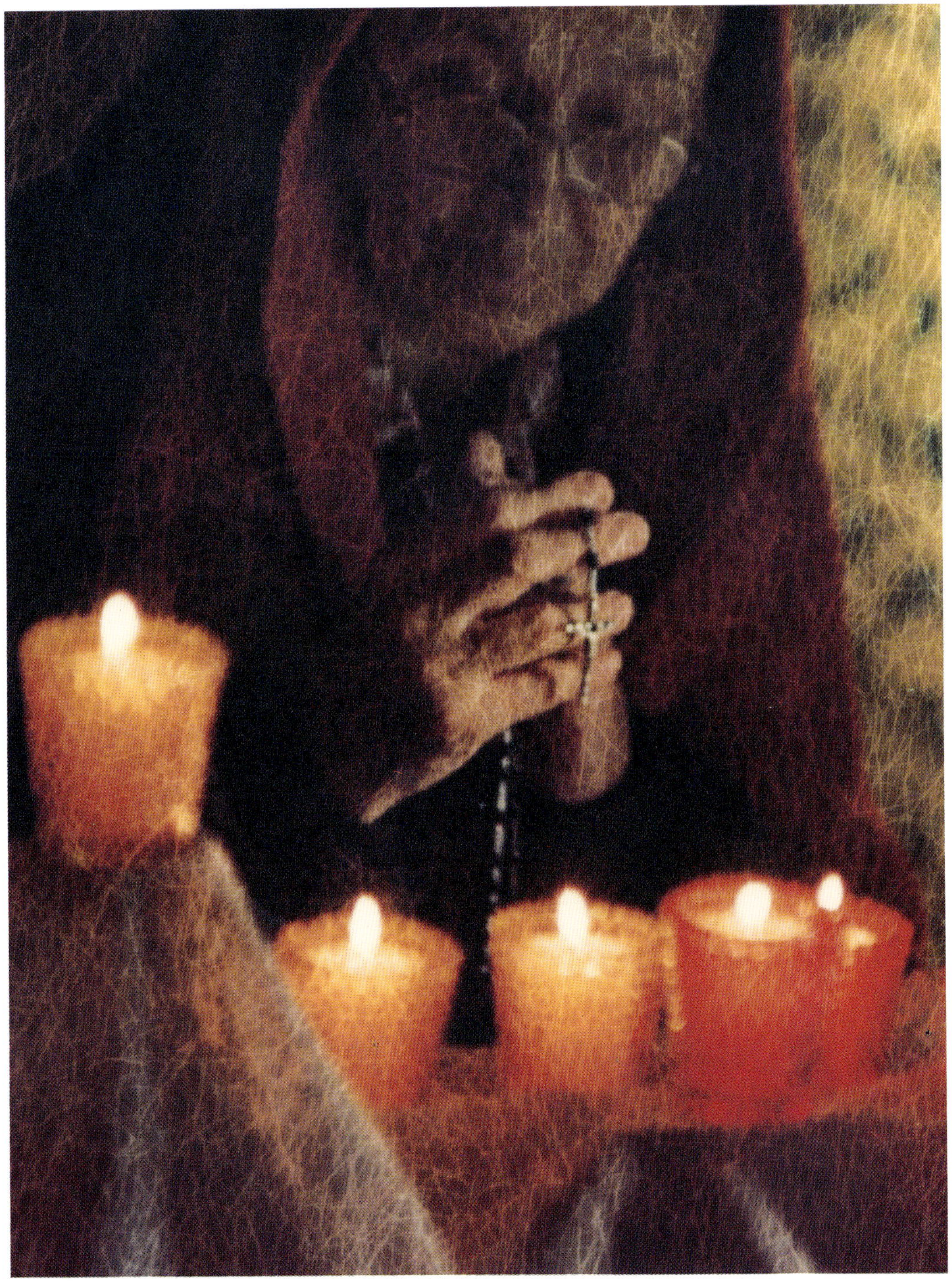

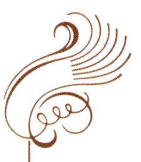

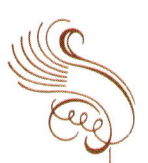

Burial of a Highlander

    The photographers searched the countryside for a hill site appropriate for this portrait. Once the site was found, they needed the assistance of members of a local Scottish Highlanders Society to complete the scene.
    The highlander on the hill played the bagpipe while the photographers worked. The music attracted farmers and neighbors who weren't quite sure what was taking place.
    "Burial of a Highlander" was taken for *Milestones*.

93

Entering the Convent

From religious rites to social rites, the photographers capture some of life's most subtly dramatic moments, such as this young woman's entry into the religious life. The wrought iron gate symbolizes the separation between the world she has chosen to leave and the world in which she has elected to live.

95

The photographers express their sincere gratitude to those people who contributed to this work: Paula Goodwin; Joann and Susan Appel; Phil Kurtz; Penny Tallio; Phil Pollard; Frankie Rogers; Brandon and Spencer White; Mr. and Mrs. Russ Foard; Rex Green; Edwin Bennett; Madeleine Benge; Laurie Clarendon; Father Migg; Mitsue Elston; Jay Marshall; Gary Ensor; Jack Foreaker; Flora Hankins Wiley; Jim Whye; Devon Robinson; Vera Hollenstein; Dorothy K. Markline; Karen Markline; Bill Silvertsen; and members of the Hartford County Highland Association.